Carbon Fiber

Paul Spencer
with Brian Yankello of Centri Designs
technical editing by Benjamin Charley

By Grace For Glory Publishing, LLC
PO Box 176
Allison Park, PA 15101

www.bygraceforglorylit.com

Cover and interior design by Brian Yankello

ISBN: 978-0-9987302-0-2

Table of Contents

Forward

Before a few months ago, I knew nothing about carbon fiber (CF).

Well, not nothing. I knew that it was strong and lightweight. I knew that it was new and kind of high tech compared to other materials. I knew that carbon fiber weaves looked cool, and I guess I just assumed that all carbon fiber looked cool because I don't think I had ever seen or noticed it in anything other than a weave. Really, I did not know much.

I now know much more about carbon fiber than the average person. I certainly still know far less than those who work in materials sciences. Perhaps I even know less than many avid car enthusiasts. But I believe through the book you've just picked up that I will be able to provide an accessible introduction to carbon fiber.

See, I have this friend, Brian Yankello, who is a bit of an entrepreneur. As long as I've known him, he has always been pushing his own limits and testing what he can know and do. He created app after mobile app in college trying to build his technical and professional expertise. And it payed off.

In 2014, Brian realized he had an opportunity in selling carbon fiber products, like bottle openers and money clips, on Amazon. After seeking out contracts with manufacturers and designing these products himself, Brian launched Centri Designs and started down the path of small business ownership. What started almost as a hobby has grown into a business that will soon eclipse all other various sources of income for Brian.

In short order, he has established himself in the industry of carbon fiber retail, and although he has much to learn in the years ahead of him, his expertise was a major player in developing this book. In fact, his presence in the industry was the motivation for the book to begin with. Without his interest in CF, mine would have never been sparked.

If you go out and look at the literature available on the topic of carbon fiber, you will notice something very obvious right off the bat. While there are many options available to you on the technical aspects of carbon fiber, from the initial booming of the industry in the 70s and 80s to the most modern techniques in production, you may be disappointed to find that there are very few resources available that provide easy access into the industry from a layman's perspective.

My hope is that this book alleviates the issue and those who find themselves causally interested in carbon fiber will use this as a resource for themselves and an introduction to an interesting topic for friends and family.

1. Introduction to Carbon Fiber

Carbon fiber has an unexpectedly common origin for what has now become the material with perhaps the single highest potential in the world of high-tech, and increasingly low-tech, applications. The birth of carbon fiber was ironically overshadowed by the whole of which it was only a part.

In November of 1879, Thomas Alva Edison applied for a patent for the "Electric Lamp," which employed carbonized organic fibers as the filaments through which the electric current passed to create the signature incandescent glow. Edison's method of producing the fibers involved dissolving cleaned cotton in a zinc chloride solution, squirting the slurry through a fluid hardener, drying the newly produced filaments, and then carbonizing the fibers out of contact with air.

U.S. Patent 223,398 - "Electric Lamp"
Edison, T.A.
Filed November 1, 1879
Granted January 27,1880

The invention consists of a light giving body of coiled carbon wires or sheets which offer resistance to the passage of electric current. The body is sealed in a glass container in a nearly perfect vacuum to prevent oxidation and injury to the conductor by the atmosphere. The current is conducted into the vacuum bulb through platinum wires sealed into the glass. Cotton threads and vegetable fibers are carbonized, coiled, and clamped to the metal lead-in wires. (Delmonte 9)

The lightbulb as a complete invention outshined these tiny fibers that were inside the glass vacuum of the bulb. And although only the electrically conductive characteristics of the fibers were appreciated at the time, within a century, the U.S. government was exploring the extent of possibilities of the fibers for industrial uses.

After these humble beginnings, carbon fiber stopped playing second fiddle when U.S. military researchers began conducting experiments for common applications in the 1950s. That decade marked the first usage of CF in the modern age. Experimentation began at the Wright-Patterson Air Force Base (Donnet 1), located just east of Dayton, Ohio. The United States military and foreign governments, notably German, Soviet, and Japanese, were those who pioneered the initial research in the field of carbon fibers. The cost of manufacturing CF at the time was prohibitively high for the private sector, but these powerful and asset-heavy institutions saw the importance of the material in advanced manufacturing.

By the 70s and 80s, the research spearheaded by public institutions began to find its home in the private sector. In spite of the still-too-high costs associated with the manufacture of CF, carbon and graphite fibers began to entrench themselves in the sector of advanced structural materials (Delmonte VII).

Koenigsegg Agera RS 'Naraya'

@miles_1995

Today, with the benefit of decades of research and development behind us, carbon fiber has overcome many of its cost concerns. Expensive, high-grade CF is still required for the most advanced and refined components of military-grade aeronautics; however, low-grade, affordable fibers are now available to the greater consumer base of western societies. We see this in the use of CF in the transportation industry, sporting goods, and medicine, as well as many other modern applications. The cost has come down so significantly that today anyone can afford to carry a CF money clip in their pocket. More adventurous consumers can design and produce their own CF products in the comfort of their own workshops and perhaps even design their own artwork using this most-modern medium. Carbon fiber still has a long way to go to realize the ubiquity that aluminum enjoys today, but as the costs continue to fall and research illuminates new manufacturing methods, we can rest assured that this lighter, stronger material will flourish in consumer goods and ultra-high-tech applications alike.

Carbon Fiber Ring

@Carbonfi

2. Carbon Fiber Basics

As we know now, carbon fiber's most important use is in advanced reinforced structural materials. Nowhere else is its impressive strength to weight ratio most well employed than in replacing large and small pieces made of concrete, metals like aluminum, or other weighty solids. The weight saving properties, which do not sacrifice structural integrity and longevity, allow for carbon fiber to be used in ways that common metals can not. And still, there is a myriad of other physical characteristics in various forms of CF that make them perfect for many applications. And all of these features that make CF great go back to the basic structure of each individual fiber - even down to the atomic level.

At the lowest level of fixed structure, carbon fiber is made up of chains of carbon bound together in repeating hexagonal shapes. Six carbon atoms come together to make one hexagon. On one side of this hexagon, a new hexagon is formed using the two carbon atoms on the first side, and a new hexagon is formed on the other side using the two carbon atoms on that side. The two atoms at the top and bottom of the hexagon can be left free, bind to other carbon hexagons creating sheets of carbon, or potentially bind to external functional groups that can alter the properties of the final fiber. The scientist producing the fibers can dictate how the atoms bind depending on the final use of the fibers.

The chains and sheets of carbon alone would leave a flat layer of carbon atoms that would be negligibly thin; however, these sheets are drawn together by weak atomic forces called van der Waals valence forces. These binding forces are considerably less strong than those between carbon atoms in a single sheet. Within the sheet, electrons are being shared between atoms which create a strong binding force. Between sheets, the atomic forces are not so direct, and thus they are more flexible.

Carbon License Plate Frame
@CentriDesigns

These sheets lay one on top of the other and bend and fold over each other in sometimes very atypical patterns. If you were to look at a cross-section of a carbon fiber, you would see something that looks similar to the end of an unsmoked cigarette. The sheets of carbon are oriented like the shreds of tobacco leaves that curve around themselves in a larger cylindrical form. Of course, carbon fiber filaments are typically much longer than they are wide which is what forms them into a longer fiber shape as opposed to the shorter cylinder of a cigarette.

With an understanding of the atomic structure of these carbon chains, sheets, and cylinders, we are now more capable of appreciating the miniscule form of a single carbon filament. Popular culture, images, and references to "carbon fiber" would have us believe that we will only find CF in woven fabrics that have been hardened with a clear epoxy into the shapes of the manufactured pieces we seek to produce. However, carbon fibers are just that: fibers. Very, very small fibers with considerably smaller diameters than a human hair.

The strength of an individual carbon fiber or filament, and thus carbon threads and carbon fabrics, is strongly related to the presence or potential for physical flaws in the fiber on the atomic level. It may not be immediately intuitive, but wider and longer carbon filaments are typically judged weaker than short, skinny fibers.

This is because increased length and diameter increases the likelihood of physical defect in the fiber. As stated by Donnet and Chand, "Almost all solids are associated with a certain degree of structural imperfection and flaws that create internal stress in the material… [Carbon fibers] are always associated with a number of flaws and defects, including dislocations, stacking faults, metal particle inclusions, gross discontinuities, or damages of the pyrolyzing fiber" (288-289).

In other words, carbon filaments can have any number of physical issues that lead to decreased strength of the entire thread. The sheets of carbon fiber may fold over each other in irregular patterns that weaken the whole. Particles other than the carbon atoms (and sought after functional groups) may remain in the fiber and cause irregularities or be removed during carbonization causing voids and discontinuities in the fibers.

Donnet and Chand went on to state that "Carbon fibers prepared under clean-room conditions showed considerably higher strengths than carbon fibers obtained under ordinary laboratory conditions… It is apparent that most of these strength limiting flaws in carbon fibers are due to contamination by impurities of the precursor fiber, which react during carbonization and heat treatment to form voids and inclusions" (297). Of course there are costs associated with keeping the manufacturing of carbon filaments as pollutant free as possible, which is one of the reasons that we see a wide range in the quality and, proportionately, the cost of carbon fiber. We will discuss this in greater length in the "Production" section.

To test a single carbon fiber may leave the observer unimpressed. A single fiber is brittle. It cannot withstand much impact or stretching before it breaks. It has a low coefficient of linear extension, and it has a high degree of anisotropy. Anisotropy is a characteristic that defines a thing as having differing properties that are directionally dependent. In other words, a carbon fiber behaves differently depending on whether you are looking down or across the thread.

@CentriDesigns

The strength of carbon fiber is derived from the orientation of the carbon chains relative to the orientation of the filament. Carbon fibers are more rigid as more of the carbon chains and sheets align down the length of the fiber. Irregularities in that structure cause weaknesses in the filament where the fiber is only as strong as "the weakest link."

Before going too far into the forms of manufactured carbon fiber, let's take a step back and look at the basic form of a composite material. "Advanced composite materials basically consist of two fundamental components: the reinforcing fiber and the matrix resin binder, which bonds them into structural members" (Delmonte VII). We have already discussed the reinforcing fiber at some length. The carbon filament, although brittle and weak as a single fiber, can be incredibly strong when appropriately associated with many other fibers of the same form. These filaments need to be drawn closely together and then bound in a solid material that will hold them in place. This binding material, or matrix, can be made of various substances, including metals, glass, plastics, and other resins. Essentially a weaker solid must be employed to hold the fibers together, and when combined, the matrix and the reinforcing fibers create a much stronger material than the sum of its parts could ever produce.

Examples of this combination are wide-ranging. Carbon fibers may be freely associated in concrete to create a stronger and lighter construction material. Likewise, the fibers may be freely associated in a synthetic resin to create durable, low-friction ball bearings. The fibers may be spun into threads which are in turn woven into fabrics. They are then impregnated with resins after being placed onto a mold to create anything from a license plate frame to a turbine blade. The threads can even be laid parallel to each other with a flexible resin to create a carbon fiber tape. The possibilities are nigh limitless.

For those who are having trouble picturing what a carbon fiber composite would look like with freely associate fibers, consider the typical form of fiberglass. It is essentially the same thing that we are talking about here, as fiberglass is also a composite material. Fiberglass, like carbon fiber, is unsurprisingly made out of fibers. Glass fibers are often bound together freely in rigid forms such as basketball hoop backboards and boat hulls. The glass fibers are generally stamped out and bound with a resin that holds them stiffly in place; however, glass fibers can be manufactured in many other ways, just as carbon fibers.

With all this in mind, consider this. A carbon fiber composite made of carbon fiber and a metal may be used in a part that previously was made completely of aluminum. The less dense carbon fiber would decrease the requisite amount of metal in the part. This would either increase the strength and durability of a part the same size, or it would allow the manufacturer to create a part of similar strength that is less massive. Either way, the manufacturer, in using an advanced composite material, is giving himself greater flexibility in the characteristics that he can produce in the finished product, probably creating a part that is more suitable for its particular use. And this part will be lighter, the benefits of which are most clearly demonstrated in the fuel efficiency gains seen in the transportation industry. Lighter is better, so long as strength is not compromised. With carbon fiber, you can have your cake and eat it too.

Koeniggsegg CCX
@Intercars

We will see more of the benefits of carbon fiber in greater detail in the following sections; however, we can wrap up this section by considering once more the benefits of this advanced composite material. Carbon fiber can add stiffness and strength to parts that were previously made of a single solid material. Carbon fiber can also reduce the thermal expansion coefficient of the composite material of which it is a part, which may be extremely important in high-tech tech applications. Carbon fiber increases the fatigue strength of the composite which will increase the durability of the parts produced. It can also decrease the electrical and thermal resistivity, which can be useful in electronic and medical applications. In total, composites offer a litany of benefits that are not available from single source manufacturing materials. To conclude this section, the words of Donnet and Chand are once again particularly appropriate: "Carbon fiber composites are ideally suited to applications where strength, stiffness, lower weight, and outstanding fatigue characteristics are critical requirements" (X11).

3. Raw Materials

You now know that the materials required for the production of carbon fiber make a very short list. Two items are needed: carbon filaments in any of various forms and a binding matrix. Of course carbon fibers are necessary; however, the fibers do not even have to truly be all-carbon. The more shortcuts in production and the less money that is spent, the less pure the resulting fibers will be. This means that a carbon fiber can end up having a host of other elements and compounds in the fiber.

True carbon fiber strands are often referred to as graphite fibers because of their high degree of purity and graphite structure. Graphite fibers have very few extraneous elements in them and are thus almost entirely carbon. These fibers are obtained by keeping strict cleanroom requirements during production and by carbonizing the organic fibers in a near vacuum to limit elemental commingling when the carbon is most susceptible to chemical reaction.

But no matter how it is produced, any carbon fiber will be primarily carbon. That is because the precursor materials used to make the thread are necessarily organic. Organic materials will contain atoms of carbon, and through the process of carbonization (explained in detail in the next section), most other elements are removed.

Among the precursors that are available for use to make carbon filaments, there are those with origins in petroleum, asphalt, coal tar, and PVC. Rayon, a synthetic thread with organic origins of its own, is also commonly used. These materials, like treated cotton threads, can be heated and carbonized into CF. They can produce much stronger threads than a basic plant precursor and are fairly inexpensive.

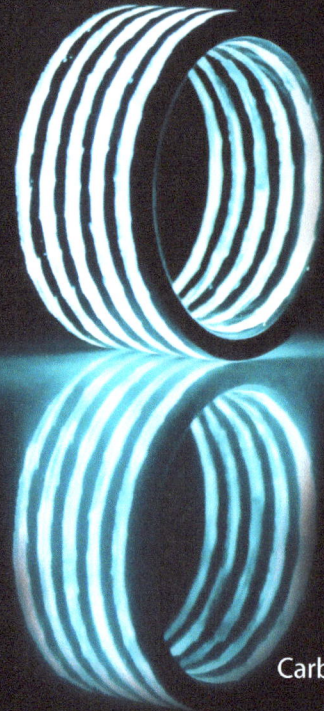

Carbon Fiber Ring
@Carbonfi

The most commonly used precursor is polyacrylonitrile (PAN). It has become the standard precursor in the industry. So long as PAN is heated slowly, it will oxidize and carbonize without melting which makes it a prime candidate for producing carbon fiber. It is a synthetic organic compound, like rayon and PVC, and can generate incredibly high yields of carbon in the fibers, unlike plant precursors. This high degree of potential purity is what makes PAN such an important precursor for modern carbon fibers.

It is important to note that in many carbon fiber components, other fibers may be employed to cut cost or generate different qualities from carbon filaments alone. Glass, metal, and plastic fibers may be spun into the carbon threads, or a completely separate fabric can be used behind the carbon fiber weave to offer a second layer of support. Of course the finished product is not what we would truly call a carbon fiber composite but would rather be a mixed composite.

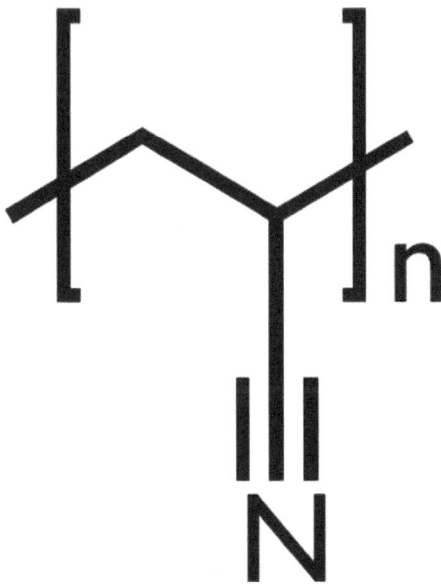

∧ PAN monomer
at the molecular level

The other half of the carbon fiber composite, often called "Carbon Fiber Reinforced Polymer," is the polymer itself; however, it is not always truly a polymer that is used. In another word, what is missing is a matrix: a binding matrix. The matrix is simply another material that will hold the carbon fibers together. Most often in today's production the matrix is a polymer. A clear plastic matrix is what binds most carbon fiber fabrics to expose the signature weaved look of the carbon threads – an attractive quality to producers and consumers alike. However, the bounds of what can be used as a matrix are very broad. The binding matrix must be first used in a liquified or non-solid form to encapsulate and impregnate the carbon fibers, and then it must return to a solid state at room temperature to effectively bind the fibers together.

So long as those requirements are met, anything could possibly be used as a matrix. Again, polymers are the most commonly used; however, glasses, metals, ceramics, vitreous enamels (powders that are melted onto the carbon fibers), and concretes are also used, depending on the final use of the manufactured part. Whatever is used must adhere to the fibers as it hardens, forming one strong and light unit of material. When the filaments are bound in the matrix, we have what we typically call "carbon fiber."

4. Production

There are a number of steps to turn an organic precursor into a carbon fiber; however, depending on how variables are controlled, the steps taken and the final quality of the fiber can differ significantly between production runs. The most basic variables are those concerned with the cleanliness of the production environment, the purity of the raw materials, and the final heat treatment temperature.

During the actual production of the fibers, the most basic need is cleanroom conditions. No dirt or dust is allowed in the production area as the raw materials are oxidized, carbonized, graphitized, and bound with the matrix. Any unwelcome particles could either change the chemical makeup of the fibers themselves or get in the way of proper fiber/matrix interaction. The more control over every particle that comes in contact with the fiber throughout the entire production process, the more the production engineer will be able to control the quality and characteristics of the final product.

Carbon Fiber Rings
@Carbonfi

The basic process for carbon fiber production is as follows: pitch preparation, spinning and drawing, stabilization / oxidation, and carbonization, optionally followed by graphitization. First, the pitch must be prepared as the production scientist desires, adding or removing any chemicals that he would like to characterize the final fiber.

The precursor must then be drawn into threads if it is not already in that form. From a pitch or other viscous liquid form, the material can be spun out and drawn into strands and then be put in contact with a fluid hardener or other chemical component to ensure that the strands will hold during the heating process.

Pagani Huayra

38

The strands are then put through an oxidation process. Oxidation is simply the process by which a material loses free electrons. Oxidation is one half of the redox reaction, in which one material loses an electron (oxidation) and another gains that electron (reduction). In the case of these carbon fibers, the precursor loses an electron to the gas, liquid (acid), or plasma elements in the environment in which it is being oxidized. This process helps to stabilize the molecules in the precursor and prepare them for carbonization; however, it can also change other characteristics of the fibers. Acids may smooth the surface and remove irregularities in the fiber, and they may also affect the fiber's overall strength, positively or negatively, depending on how the strength of the final thread is measured. Without diving into great detail, we can rest assured that those engineers producing the threads employ different chemicals throughout production to change the final qualities of the carbon fiber.

After oxidation, the precursor threads are heated to initiate the carbonization process. To best explain carbonization, think about a bonfire. Wood in a bonfire is burned and the ash left over in the fire pit is mostly carbon. However, the material in a fire has combusted, and in the carbonization of carbon fibers, combustion is specifically avoided. At high temperatures, without flame, extraneous molecules are removed from organic material leaving the carbon behind in higher concentrations. The higher the concentration, the purer the carbon. At very high temperatures, after the non-carbon elements have been removed, about 50% of the original PAN will remain. In the whole production process of carbon fiber, "the mechanical properties… can be enhanced by varying the process parameters, the most important of which is the final heat treatment temperature, and by modification of the precursor fiber materials" as we've already discussed (Donnet 8).

The heat treatment temperature (HTT) and the rate of heating are incredibly important variables in the production of carbon fibers. If the heating rate is too high, the precursor may lose too much material, but low rates of heating involve more time, and thus, more cost.

The final HTT is one of the most important details of the production process because it will help determine the purity and directly affects the molecular structure of the fiber. The higher the temperature at the end of the heating process, the more pure and organized the resulting carbon will be. The structure of the carbon will be more uniform with fewer incongruities. This purity of form and composition increases the quality of the fibers. The carbon hexagons align more uniformly along the length of the fiber, and this orientation also increases the overall rigidity of the fiber. This highly organized graphite fiber differs from its basic carbon fiber cousin in various forms of measured strength and other characteristics. Carbon fiber, depending on the final HTT and precursor used, can vary significantly in form, function, characteristics, and final use.

When we are talking about heat treatment temperature and final quality of threads, we can actually organize the carbon fibers into three different loosely applied types. Depending on the scientist and industry, the categorization of individual filaments may differ, but the types are a useful tool for understanding different quality threads.

Type III is obtained at the lowest final HTT, or even at high temperatures when using a PAN precursor. These turbostratic fibers are made of randomly organized carbon sheets and have high tensile strength. They are less expensive to make than the other two types. Type I fibers are highly graphitized (highly organized) and made from a mesophase pitch precursor. They resist stretching under load and have high thermal conductivity. Type I fibers have higher HTT and cleanroom requirements and are thus more expensive to produce than the other two types. Type II fibers are somewhat organized on the molecular level and share characteristics of the two other types. They are essentially a compromise between Type I and Type III and have their own distinct qualities, which are a blend of the qualities of the other types. Each type has its own strengths and weaknesses and can be used in different ways in the same broader application. Although Type I may be required for certain parts in a spacecraft, Type III may be usable and even preferable in other parts, depending on the characteristics needed for individual components.

Koenigsegg Agera R

@Intercars

The graphitization of carbon fibers can almost be considered a substep involved in the carbonization process. As the heating of the precursor threads increases, extraneous materials are removed, and the carbon atoms become more and more highly organized; they become graphitized. Graphite has a crystalline organization of carbon atoms as do the carbon fibers which are heated to extremely high temperatures. We sometimes even call these filaments graphite fibers rather than carbon fibers. "The distinction between carbon fibers and graphite fibers is primarily one of degree of crystalline order. Physical differences, such as density of fibers and elastic moduli, will also be apparent" (Delmonte 48). Truly, the materials are not formed into a genuine graphite structure. In fact, such a high degree of crystalline order may be avoided. But the final CF will be of high order and high quality if heated to a high temperature.

Another factor that can greatly affect the final product is stretching during the heating stage. "Application of external stretching load at high temperatures induces intensive cleavage of intermolecular bonds, which otherwise interfere with intra- and interfiber alignments, resulting in better ordering of the structure" (Ermolenko 37). The stretching removes bonds that would otherwise detract from the strength of the fibers while also helping to align the ordering of the molecular bonds along the length of the fiber, strengthening it.

Something else to consider at a molecular level is the presence of gases and spray-on liquids that can be used as a surface treatment for the fibers. These gases, as alluded to before, can change the characteristics of the surface of the fiber by adding functional groups to the ends of the carbon sheets. These functional groups are an incredibly important detail when we consider how a matrix will bind to the fibers. Some matrices will not naturally bind to carbon fibers because of the incongruence of molecular interaction; however, by adding other elements to the carbon sheets, a matrix will become more receptive to bonding than it otherwise would have been. Because of the breadth of possible functional groups and matrices, we will not pursue any details here, but once again, this is one of many variables that the manufacturer can alter which will change the characteristics of the final product.

Mclaren P1 GTR

@Intercars

@Carbonwaves

After the manufacturer obtains the final carbon filaments, they are sometimes chopped into short pieces. They can be applied to a matrix without any other steps which creates a free association of fibers suspended in the solid matrix. But often, the chopped fibers are mixed with liquid to form a slurry of fibers that is spun at high speeds in a rotating tube to create yarns of carbon fiber. The CF can again be used in this form; however, often the yarns are woven into fabrics which can then be bent into countless shapes before a matrix is applied and hardened. The variety of manufactured pieces possible from this point are countless. The creativity of CF manufacturers has given the 21st century a broad spectrum of CF goods, which we will explore on the coming pages.

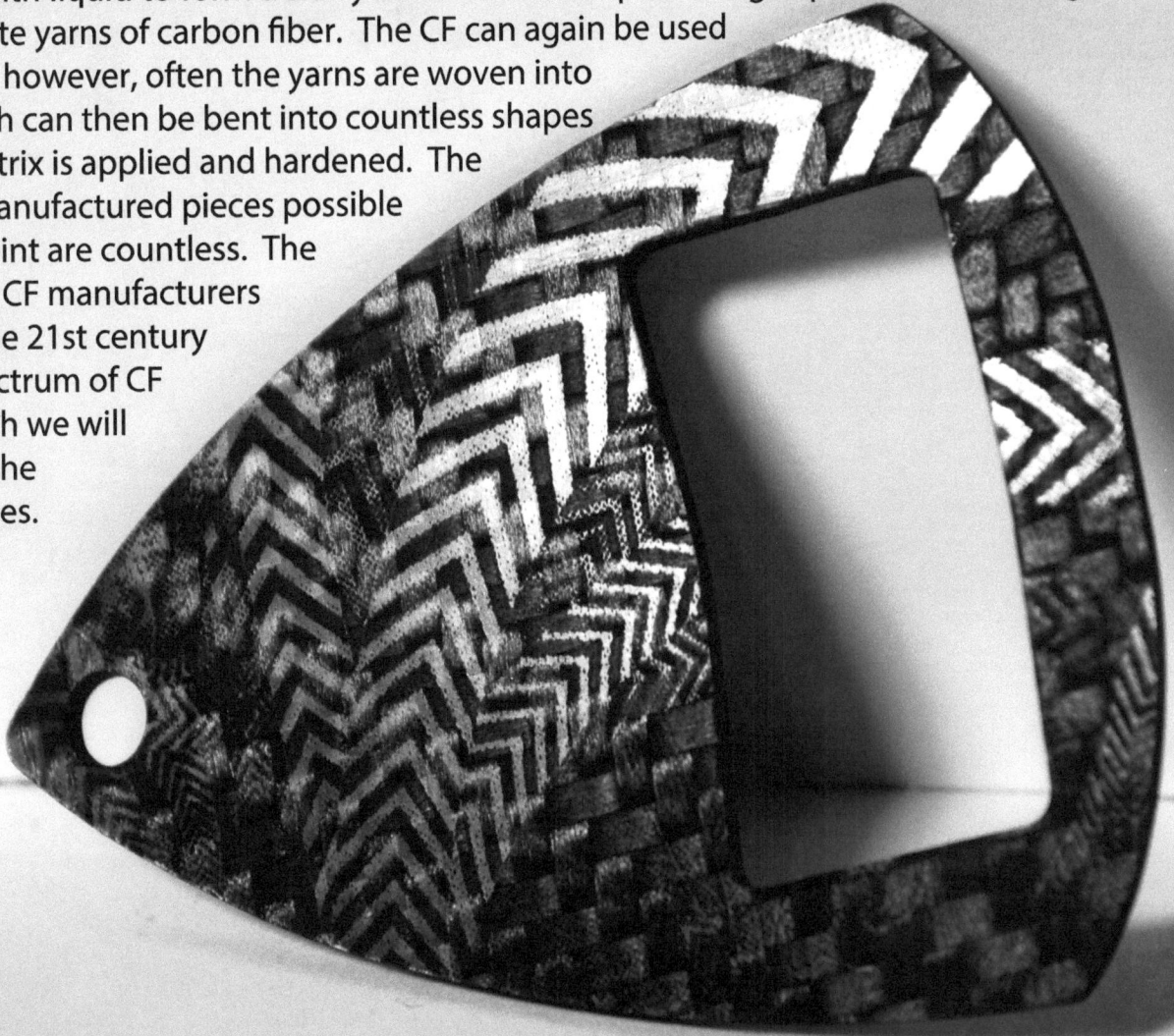

Lamborghini Huracán - Forged Carbon Fiber Engine Cover

5. Applications

Much like when we think about the fibers themselves, when considering the applications of CF, we must divide the topic into two loose classifications. We have discussed the difference between Type I and Type III fibers and their various qualities. Type I fibers have fewer imperfections, which carries a host of benefits over their lower quality counterparts, but sometimes the qualities of Type III fibers are preferred. As stated previously, the two different types can even be used in the same general application for different purposes.

Carbon Fiber Bike Frame

This distinction in quality is what divides CF into two main categories when applied to manufactured goods. "The two main sectors of carbon fiber applications are the high technology sector, which includes aerospace and nuclear engineering, and the general manufacturing and transportation sectors, which include engineering components such as bearings, gears, cams, fan blades, etc., and automobile bodies. However, the requirements of the two sectors are fundamentally different" (Donnet 370).

Although car enthusiasts may not like to hear it, automotive and transportation applications of carbon fiber do not usually require the same quality as spacecraft applications, though Type I fibers may be used for supply chain convenience or marketing purposes. Lower quality and turbostratic CF is acceptable for automotive components and the body of the cars. It is strong enough to hold form over time, support heavier components like the engine, and take significant impact. Conversely, CF used to make particular components in spacecraft must be able to withstand intense stresses without any imperfections that would lead to cracks or breaks in the material, which could lead to catastrophic failure.

Ferrari LaFerrari

@Intercars

As expected, one of the very first industries to embrace the use of carbon fiber was the automotive industry, and in particular, automobile racing. The first car to contain carbon fiber in the construction was the McLaren MP4/1 in the 1981 Formula 1 season. McLaren first used carbon fiber in what is called a composite monocoque which is now common in racing and street-legal supercars. Monocoque is a French term, meaning "single shell." The carbon fiber monocoque is a single surface made of various types of carbon fiber fabrics oriented in calculated directions to provide the perfect structure and frame for the strength of the vehicle. The rigid and lightweight chassis allows for optimal acceleration, braking, and predictable handling.

The carbon fiber used throughout today's Formula 1 cars not only increases the performance of the cars but can also provide additional safety for the drivers. The monocoques are extremely safe and form a strong shell that can help prevent shrapnel from other vehicles from coming in contact with the driver. A testament to this added safety is when Fernando Alonzo crashed into a wall going 193 MPH at the Australian Grand Prix in 2016, disintegrating his car. Amazingly, Alonzo was able to walk away from the crash.

Carbon fiber has become one of the leading materials in the automotive industry. Nearly all supercars, hypercars, and modern sports cars incorporate CF in some aspect. Manufacturers like Pagani and Koenigsegg are among the leaders in making carbon fiber the primary material for the entire vehicle. From the rims to the sway bars, they utilize both the strength and weight properties to make the vehicles go faster and handle better than anything conceived before them. It has become a ubiquitous material that has not only defined modern performance but also offers stunning looks. As costs decrease, tuners are now using carbon fiber for aftermarket performance parts due to its aesthetic nature. The craftsmanship that goes into making perfect, void-free carbon fiber parts for cars is very much admired by automotive enthusiasts.

@miles_1995
Koenigsegg Agera RS 'Naraya

Lamborghini is one of the major new players utilizing "forged carbon fiber" in their vehicles for parts that have otherwise been impossible to manufacture using CF. Forged composites use a patch of fibers (roughly 500,000 turbostratic fibers per square inch) mixed with resin that is then put under high compression to stamp out pieces in nearly any molded shape imaginable. This variation of carbon fiber allows for the creation of new shapes not possible from the traditional woven fabric due to the fabric's physical limitations. While this is revolutionary in respect to the variety of products now available for production, forged carbon fiber is not proven to provide the inherent strength woven or unidirectional composites offer; however, forged carbon fiber allows for significant cost and time savings compared to previous methods of production. This technological breakthrough may allow the everyday sedan to utilize carbon fiber in its construction to help reduce weight and ultimately improve the car's fuel efficiency.

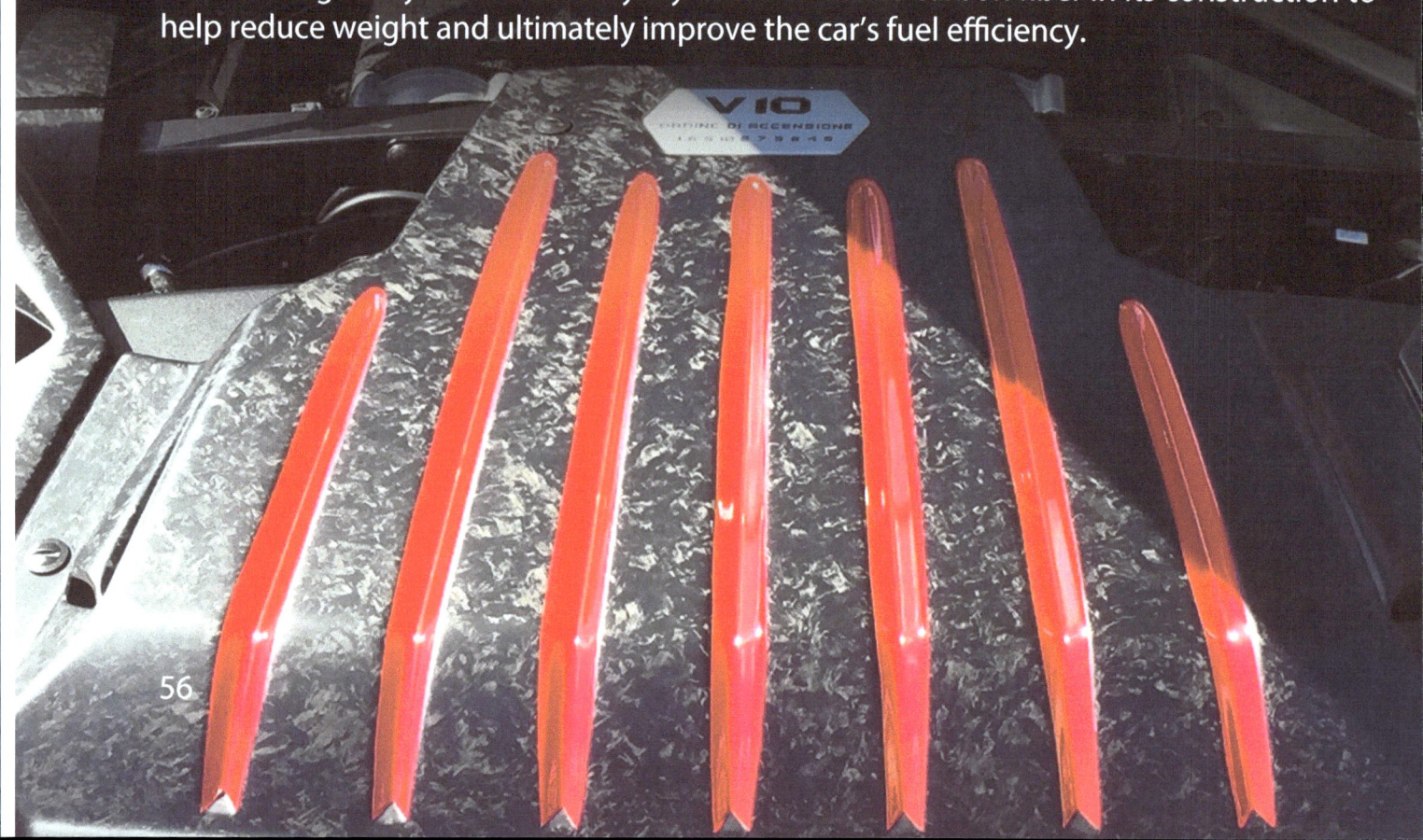

Some other basic applications for low quality CF are those for everyday use. Golf clubs, ski poles, lacrosse and hockey sticks, tennis racquets, and baseball bats, among other sporting goods equipment sometimes employ carbon fiber in the shafts to keep the weight down, increasing performance while not sacrificing any strength over aluminum or other metallic alloys. Sports and protective helmets also employ CF for their shells as it adds support to that which would otherwise be simply plastic. In fact, the MLB has mandated that all helmits incorporate carbon fiber to withstand 100+ mph pitches. By infusing the plastic matrix with carbon fibers, they will dramatically increase its resilience and firmness. Carbon fiber also has incredible resistance to wear. It can be used to produce bearings and brakes that last much longer than the metal parts that they are increasingly replacing.

Like in other common applications, the CF used in construction materials, those incorporated into concrete, will be of the low-modulus low-quality variety. CF produced in the large quantities required for gross applications are restricted to the more economically reasonable methods of production, which in turn limit the quality of the final product. Carbon fabrics, fibers, or particles can be used in the concrete mix, and depending on how they are used, the fibers can help control the electrical conductance of the hardened concrete. This may be important in the design of buildings at risk for electrical shock from lightning. The added carbon fibers can also add strength to concrete that will remain in the presence of corrosive materials. The atomically inert qualities of CF slow the damage done by the natural corrosion of chemicals.

Carbon fibers are electrically conductive which means they can be used for electronic purposes. They also conduct heat very well. Combining these qualities, CF tapes and fabrics can be used where conduction and flexibility is important. The elements can be used to warm fabrics and blankets, and they can heat pipelines to avoid freezing and bursting. CF can be used to control the temperatures of houses, barns, and pressurized storage tanks. The material can be incorporated into sidewalks, driveways, and airplane landing strips to melt hazardous ice or potentially to evaporate excess water.

Carbon Fiber Ring

@Carbonfi

Northrop B-2 'Stealth Bomber'

Although CF conducts heat extremely well, it is also known for its low and controlled thermal expansion. This means that as the material changes temperature, sometimes to very high or low extremes, it does not expand, warp, or otherwise change shape or size. This is an important quality when it comes to precision in extreme environments. When objects are launched into space, they necessarily heat up while shooting out of the atmosphere and then cool way down when subjected to the vacuum of space. For telescopes made to detect electromagnetic radiation in space, from radio waves to visible light to gamma rays, it is important for their components to remain in their precise places – absolutely and relative to themselves. If the lenses in the light telescope or the antennae on a radio telescope were to shift at all, the measurements and observations from the telescope would be inaccurate and unusable. The low thermal expansion of CF holds the casing and antennae of the telescopes precisely in place to avoid those issues.

Carbon fiber, due to its electrical qualities, is also useful in a number of applications in electronics. It can effectively block electromagnetic fields and so can be used for housing or coating electronics that are sensitive to the fields created by other electronics and their wires. CF essentially dampens the static present in an environment, so that sensitive electronics can operate without interference.

CF can also act as a phenomenal chemical filter. CF adsorbs chemicals, which means that the atoms of the chemical stick to the surface of the carbon fibers. This differs from absorption by which the chemicals would be drawn into the carbon fibers. When the surface of carbon fiber is activated by adding specific functional groups to the carbon fiber sheets, the adsorption can be dramatically increased. In manufacturing settings, this CF can be used to filter out harmful chemicals from air or water before it is reintroduced to the environment, greatly decreasing the pollutants in nature. Also, through other chemical processes, these adsorbed elements can be cleaned off of the carbon fiber. This means CF can be rinsed and reused many times before it is no longer usable for these purposes.

Carbon Fiber Ring
@Carbonfi

In the same way, carbon fiber sheets can be used in biological and medical applications. Carbon fiber can filter out harmful chemicals from a patient's bloodstream, which can be more effective and less invasive than other methods. It can also be used to filter out solvents added to solutions in laboratory and manufacturing settings. This allows scientists to remove and recycle expensive or harmful solvents that would otherwise be lost in the useless wastes that are disposed of after experimentation and production.

Another application that we will consider for CF are those in radioactive settings. Carbon fibers are highly stable in the presence of radiation which "makes it possible to use element-carbon fibers [e.g., graphite fibers] in technological processes dealing with radioactive substances, as well as to adsorb radioactive compounds that may further be used as sources of radiation" (Ref H, 246). Carbon fiber can withstand the punishment from radiation and can even (like with the useful solvents in solution) save radioactive material that can be reused instead of being thrown out with waste material. CF is also transparent to X-ray radiation which makes it useful in some medical applications. CF can be used to make the wires, tables, and other components in X-ray rooms so that doctors do not have to unnecessarily move and shift patients to get every angle needed to be photographed. Aluminum, other metals, and many composite materials cannot offer this kind of flexibility.

Lastly, although CF can be made to be an excellent adsorber, as we have mentioned, it is chemically inert. This makes it very useful in biological applications. Where metals may corrode, oxidize, and rust, CF maintains its inactive chemical composition. This makes it a prime candidate for prosthetics and replacement bones and joints. The strength of the material ensures the strength of the replaced body part while also ensuring that the patient will have no negative reactions to the material itself.

Mclaren P1 (Front)
Ferrari 458 Speciale (Back)

In some ways, it may seem that carbon fiber is some sort of miracle material. It certainly does have some very intriguing qualities that makes it especially useful in many different applications. But when we go back to the basics of CF, we see a strong, inflexible fiber that can add strength and lower weight in composite materials. Although the fibers themselves are brittle, in threads and fabrics, they compound their strength and minimize their weaknesses. And when incorporated into metallic and polymer matrices, the weaknesses become less and less noticeable as the strengths rush to the forefront. Carbon fiber, when mixed with aluminum, titanium, beryllium, glass, boron, plastics, and rubbers, becomes an incredible material with countless practical uses.

And for those less practical uses, carbon fiber simply looks really cool. Go check it out for yourself. There are some great, unique products out there for sale at reasonable prices. And if you really get into it, you can make your own with kits from a number of producers. Have fun exploring the breadth of what you can do with Carbon Fiber.

Special Thanks

Brian Yankello
CentriDesigns.com | @CentriDesigns

Benjamin Charley
Technical Editing

Juan Martinez
@BeersandCameras

Yannick Schaller
@Intercars

Gerco Marsch
CarbonWaves.etsy.com | @TurboCarbon

Christian Morris
@christianalexandermorris

Tyler Wright
Carbonfi.com | @CarbonFi

Miles Toussaint
@Miles_1995

Without the valuable perspectives of the three following people, this book would have never been produced. With their knowledge and innovative spirits we have the chance to see carbon fiber expand into untapped applications, industries, and arts.

In the following pages I intend to give deeper insight into what inspried them to consider and pursue carbon fiber. I hope their stories will inspire you to do the same.

Brian Yankello
Owner - Centri Designs
Retailer

Brian Yankello has always been into cars, especially high-end racing cars. The first car that caught Yankello's attention was the Ferrari Enzo, and that car had carbon fiber in it. This was how Yankello was introduced to CF, and since then, he has been intrigued by the material. As the cost decreased and CF became more accessible, more cars incorporated it. Because of the greater accessibility, Yankello started to see and appreciate these cars and carbon fiber in person, which only increased his interest.

Throughout his life, Yankello has been artistic, always finding himself interested in new materials. He learned his art by painting and working wood, among other mediums. When his interest in art and carbon fiber collided, he did some research on how to actually make CF pieces. He laid out each step and found that he would have up front costs of about $250, so he swallowed the expense and started experimenting with CF weaves. As he continued to work with the material, Yankello found that although anyone can reasonably make a decent CF piece, it takes a lot of practice and great attention to detail to make a piece void-free and of high quality. Because of that required precision, Yankello has great respect for the men and women that produce high quality pieces.

Within the last 5 years or so, CF has finally become inexpensive enough for people to pick up making CF pieces as a hobby and for parts to be readily accessible at a reasonable price. Seeing this, Yankello started his own company, Centri Designs, in 2014 with the goal of making products that everyone could afford. Yankello sought to introduce CF to a broader audience beyond the car enthusiasts who have become intimately familiar with the material over the years. Through his products, blog, and his support of this book, Yankello sees his role in the CF industry as being a liaison between a highly technical, niche market and the greater public. As the years pass since the founding of his company, Yankello continues to broaden his reach and broadcast his message: Carbon Fiber is for everyone.

Centri Designs

Gerco Marsch
Owner - CarbonWaves
Manufacturer

Gerco Marsch first came into contact with carbon fiber through his RC (radio controlled) hobby as a teenager in the Netherlands. After being taught all the tips and tricks of composites from an older member of the RC club, Marsch found himself especially prepared to incorporate the new material of carbon fiber into his RC planes.

Over the 25 years since those early days, Marsch has developed an expertise in carbon fiber manufacturing. When he started to work in aeronautical engineering, Marsch worked in labs to produce the highest quality CF he could. Marsch was initially familiar with the basics of wet layup methods and a technique he developed incorporating a water balloon to push out voids in the CF. As his experience in the lab grew and new techniques came to light, Marsch began to use the more modern methods of resin infusion and vacuum bagging; however, the pursuit of a perfect, void-free CF piece is still as arduous as ever.

Now as the owner of CarbonWaves, Marsch strives to produce the best in luxury carbon fiber materials. His company produces high-quality carbon fiber plates while incorporating a creativity that makes his products stand out from the rest of the industry. He uses materials of different colors and metal threads in his weaves to produce unique final products that push the bounds of what CF is and can be.

With the cost of materials coming down and the cost of production coming down even faster, Marsch believes that CF will become more and more accessible and ubiquitous beyond the luxury goods market. He admits that in spite of his company's creativity in producing unique CF weaves, he has barely scratched the surface of what is possible with CF. Marsch is excited to be in a position to witness and influence the great strides and changes that are sure to characterize the carbon fiber industry in the years to come.

Christian Morris
Artist

Christian Morris is an artist, plain and simple. When he looks at the world around him, he is constantly looking for new perspectives, new forms, and new mediums. He demonstrates skill in sculpting using various materials, and as a student, he loves the opportunity to be pushed and challenged with interesting assignments. Morris has a very expansive view of art and is not particularly exclusive on what he would accept as art by other artists. He is a man who realizes that art has an incredible quality of being widely accessible and infinitely flexible.

Morris understands that each artist brings his or her own backgrounds and interests into play when he or she creates the art. So when he was asked whether he saw carbon fiber as a viable artistic medium, he said, "Of course." Artists, like us all, have interests outside of their work, and those with aerospace and auto racing backgrounds would naturally be inclined to know about CF and to use it in their own works.

Carbon fiber, as it now stands, has not been introduced to the arts community in a meaningful way. Very few artists are aware of the availability of the material, and fewer are using it. But Morris believes that the lack of exposure of CF as a medium is its primary barrier to entry into the arts community. No one has truly introduced the material to artists even though it has great aesthetic qualities as well as positive structural attributes for use in large, freestanding installations.

Morris believes that artists simply need to be made aware of the material, how it can be used, and how it is made. A simple how-to video or a workshop in a classroom could open a world of possibilities to artists who could find a multitude of uses for CF. Artists just need to see a material worked with once, and after some trial and error, we could expect to see their boundless creativity and skillful hands build CF pieces that would seem very exotic compared to its common industrial uses. Morris says, "People are afraid to try new things in general, but artists, not as much." Maybe it will be artists themselves who are passed the torch from passionate individuals like Yankello to truly introduce carbon fiber to the rest of the world. Morris, with his connections to Fiberart International, an organization that promotes art that employs fibrous materials, is in a unique position to be this harbinger of change.

References and Suggested Resources

"Carbon Fiber." Fibre Glast. Fibre Glast Development Corp., n.d. Web. 6 Feb. 2017.
    <http://www.fibreglast.com/category/Carbon_Fiber>.
"Carbon Fiber News and Information." Autoblog. AOL Inc., n.d. Web. 6 Feb. 2017.
    <http://www.autoblog.com/tag/carbon+fiber/>.
"Composites Today." Carbon Fibre News. Composites Media Ltd., n.d. Web. 6 Feb. 2017.
    <http://www.compositestoday.com/tag/carbon-fibre/>.
Delmonte, John. Technology of Carbon and Graphite Fiber Composites. New York: Van Nostrand Rein
    hold, 1981. Print.
Donnet, Jean-Baptiste, and Roop Bansal Chand. Carbon Fibers. New York: M. Dekker, 1990. Print.
    Ermolenko, Igor Nikolaevic., Nina Gul'ko Vladimirovna., I. Lyubliner P., N. Gulko V., I. Ermaolenko
    N., Ilja Ljubliner Petrovic, and Titovec P. Chemically Modified Carbon Fibers and Their Applica
    tions. Weinheim: VCH, 1990. Print.
Gill, Richard Malcolm. Carbon Fibres in Composite Materials. London: Iliffe for the Plastics Institute,
    1972. Print.
"How Is It Made?" ZOLTEK™ Carbon Fiber, n.d. Web. 6 Feb. 2017. <http://zoltek.com/carbonfiber/how-
    is-it-made/>.
"Materials / Tools." Carbon Fiber Raw Material. Rock West Composites, n.d. Web. 6 Feb. 2017.
    <https://www.rockwestcomposites.com/materials-tools>.
McConnell, Vicki. "The Making of Carbon Fiber." CompositesWorld, 18 Dec. 2008. Web. 6 Feb. 2017.
    <http://www.compositesworld.com/blog/post/the-making-of-carbon-fiber>.
"An Inside Look at BMW's Carbon Fiber Manufacturing Process." MotoringFile. YouTube, 01 Aug. 2011.
    Web. 06 Feb. 2017. <https://www.youtube.com/watch?v=kaoq8Mc4xxw&t=617s>.
"Online Composites Learning Area." Easy Composites, n.d. Web. 6 Feb. 2017.
    <http://www.easycomposites.co.uk/#!/composites-tutorials>.
Park, Soo-Jin. Carbon Fibers. New York: Springer, 2015. Print.
Soller, Jon. "Weave Definitions." Soller Composites, n.d. Web. 6 Feb. 2017.
    <http://www.sollercomposites.com/fabricchoice.html>.

www.ingramcontent.com/pod-product-compliance
Lightning Source LLC
Chambersburg PA
CBHW051601190326
41458CB00030B/6499